THE SCALE OF THINGS

MIKE FAIRBRASS
DAVID TANGUY

quadrille

Publishing director: Sarah Lavelle
Creative director: Helen Lewis
Design & Art Direction: Praline
(David Tanguy & Giovanni Pamio)
Words: Mike Fairbrass
Editor: Harriet Butt
Production: Vincent Smith, Nikolaus Ginelli

Quadrille is an imprint of Hardie Grant
www.hardiegrant.com

First published in 2017 by
Quadrille Publishing Limited
Pentagon House
52–54 Southwark Street
London SE1 1UN
www.quadrille.com

Cataloguing in Publication Data: A catalogue record
for this book is available from the British Library.

ISBN 978 1 78713 057 9
Printed in China

CONTENTS

FOREWORD

Architecture is about solving problems in three dimensions, and from its earliest days, scale models have been a tool to explore and explain possibilities, considering in the studio what will one day be built in the city.

I have never ceased to delight in the miniature worlds conjured up by architectural models and I love the feeling of human scale experienced when entering a built space for the first time, or the sense of being one human in relation to millions in our vast modern cities.

Through all the years that Mike Fairbrass led the modeling team at my practice, he always brought to the task the sense of engagement and delight in the challenges of rendering structures at small scale. The same playful yet thoughtful approach is visible in Mike and Dav's book, which uses scale and juxtaposition to illuminate complex sociological issues of power and consumption. Now more than ever we should all be reminded of our position and influence in relation to our increasingly complex world, its environment and its relationship to the cosmos. Understanding scale makes us better citizens.

It is therefore a genuine pleasure for me to recommend the pages that follow.

Richard Rogers

INTRODUCTION

NOUN

1. Plates protecting the skin of fish and reptiles.
2. An instrument for weighing,
3. The relative size or extent of something.
   (Last in the dictionary but by far the most interesting.)

THE FERMI TECHNIQUE

This book uses readily available common knowledge from a wide range of sources transposed into different Scales. To achieve this we employ the Fermi technique named after physicist Enrico Fermi who pioneered a method to determine approximate answers using justified assumptions about difficult to calculate quantities.

"On a cosmic scale, our life is insignificant, yet this brief period when we appear in the world is the time in which all meaningful questions arise."
Paul Ricoeur

THE SUN'S AVERAGE DISTANCE FROM EARTH IS AROUND 93,000,000 MILES, BUT WHO CAN CONCEIVE OF THAT DISTANCE? USING SCALE THE SUN BECOMES A GRAPEFRUIT, THE EARTH A GRAIN OF SAND AND THE DISTANCE BETWEEN THE TWO A LONDON BUS.

Suddenly it fits in your head.

We discovered a shared fascination with scale when whilst working on a touring exhibition for architect Richard Rogers (Mike responsible for the scale models and David in charge of the graphics). We realized you can use scale as a tool not only to compare but to influence our perception and 'feel' new concepts – to render the unimaginable imaginable so that we may further understand and appreciate our world and its universe.

Our ambition for this book is to entertain, inform and engage you with amazing facts morphed by scale and graphically illustrated. From the deepest ocean trench to all the stars in our galaxy via T. Rex's shoe size, we will take you beyond distance to explore height and depth, molecular and atomic size, the scale of wealth, the speed of thought and plate tectonics, the power of bombs or dictators and the weight of the internet.

The Scale of Things is a book for those who want to go further than just information. We want to reach into your mind and make bits of it boggle as you picture and percolate the joys of science and understanding conveyed in a new way.

The human scale is 1:1, our point of reference. Let's go beyond...

Mike Fairbrass & David Tanguy

BIOL

OGY

If you removed the empty space from their atoms, the human race would occupy the volume of a single sugar cube.

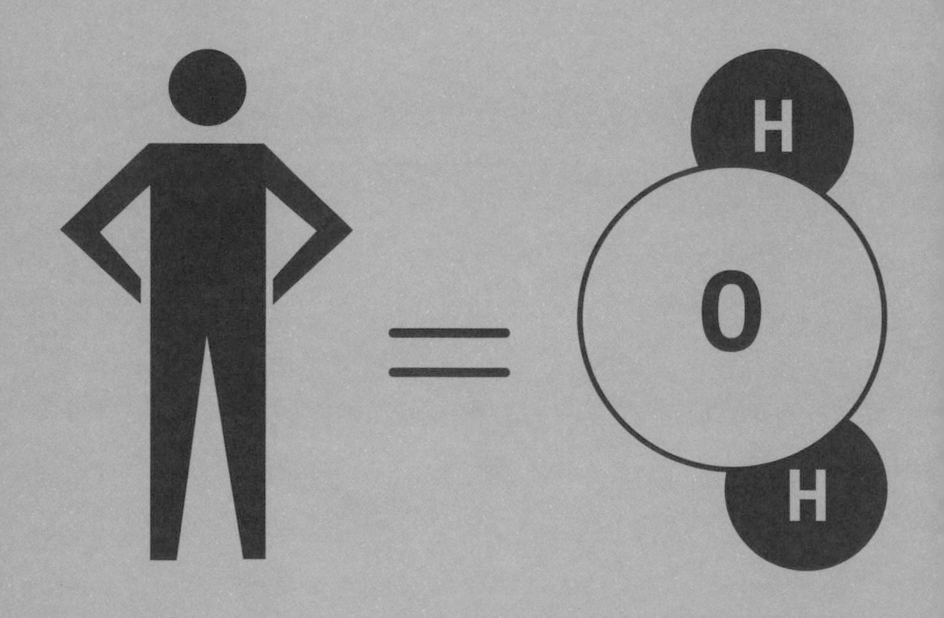

If a water molecule was
the size of a person,

a raindrop would be bigger
than the Earth.

The average office desk has as much bacteria as four hundred toilets.

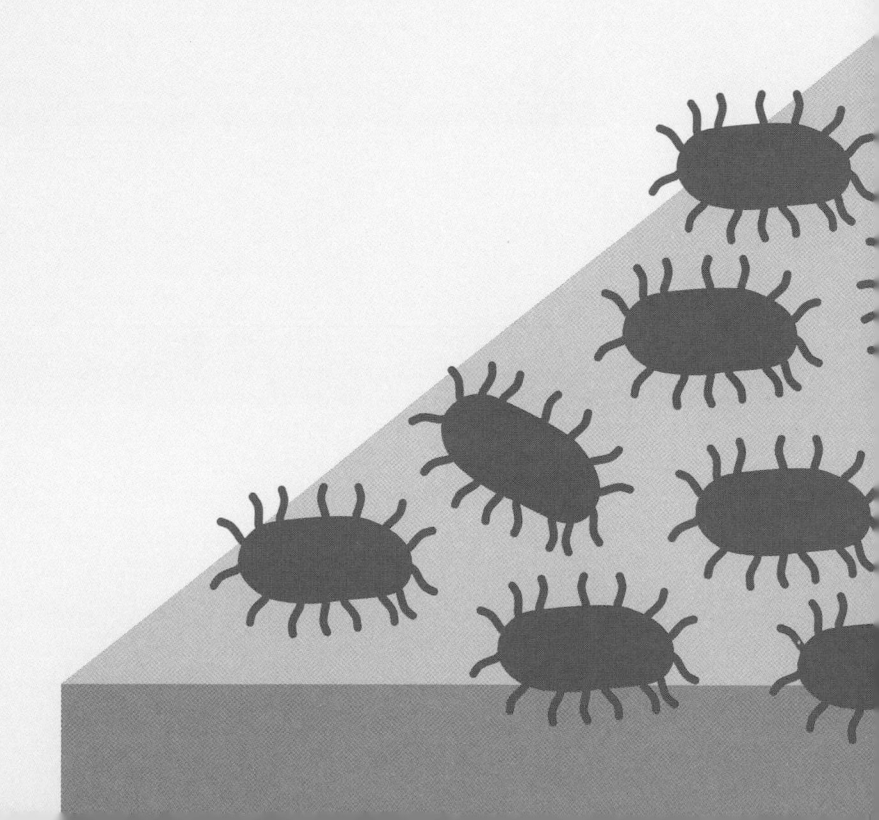

If an atom was the size of a typical city block,

its nucleus would only be the size of a football.

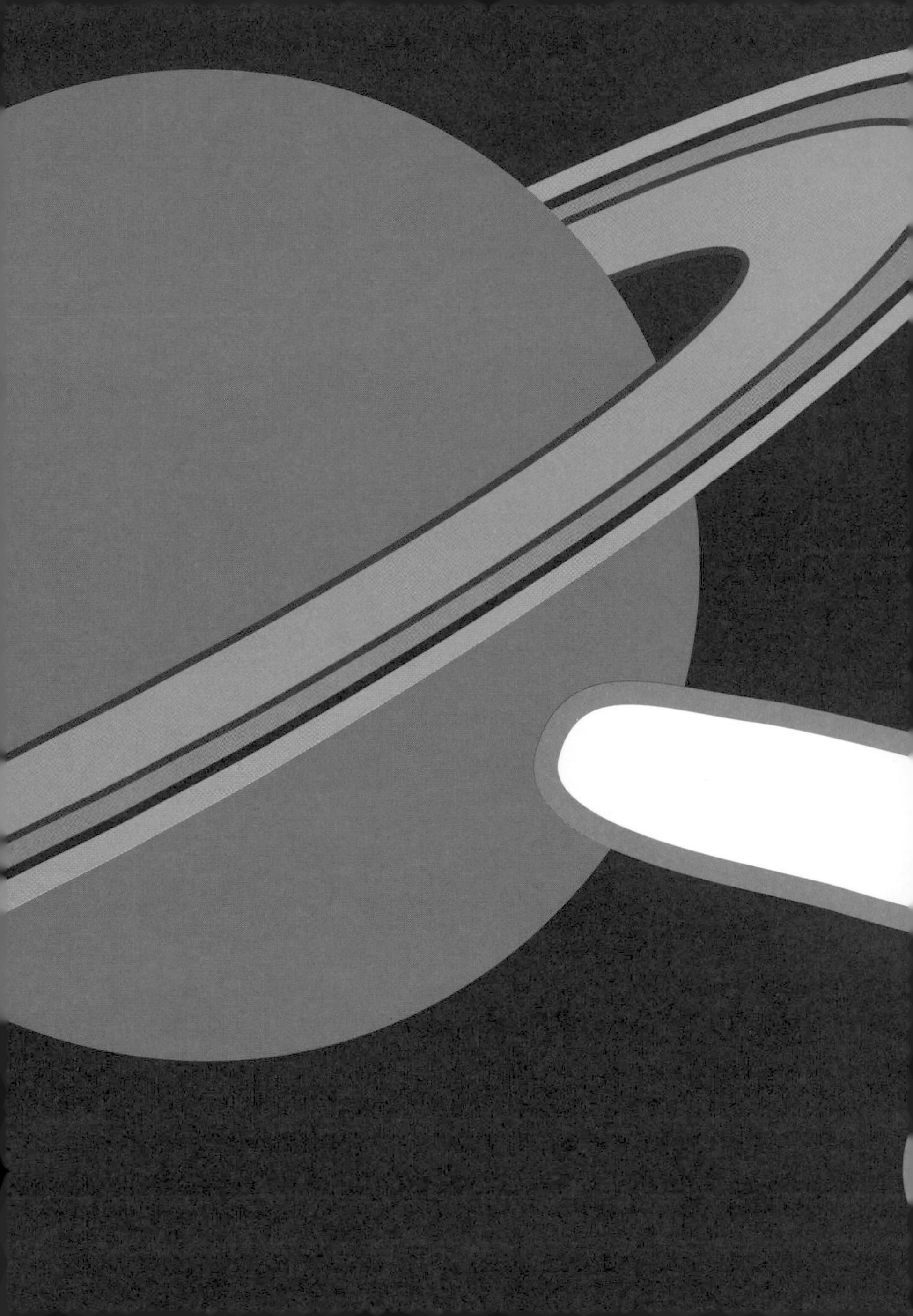

If a human cell was the size of a tennis court, a single virus would be the size of a golf ball.

At that scale a person would be tall enough to reach out and touch Saturn.

You have as many brain cells as there are trees in the Amazon rainforest and more synaptic connections than all the leaves.

The total length of the brain's neural pathways...

...is roughly the distance
to the moon.

COS

MOS

If the sun is the size of a grape-fruit, the earth is a grain of sand, and the distance between the two is a London bus.

Grapefruit / Sun

Grain of sand / Earth

Matter from a neutron star is so dense...

...a teaspoonful of it would weigh more than the human race.

In the time it takes a fly's wing to flap,

light has travelled from London to Hamburg.

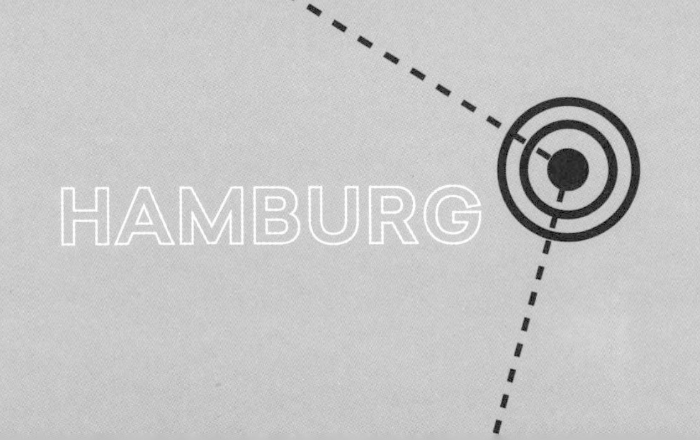

If each of the Milky Way's stars were the size of a pea and you poured them into an Olympic sports stadium...

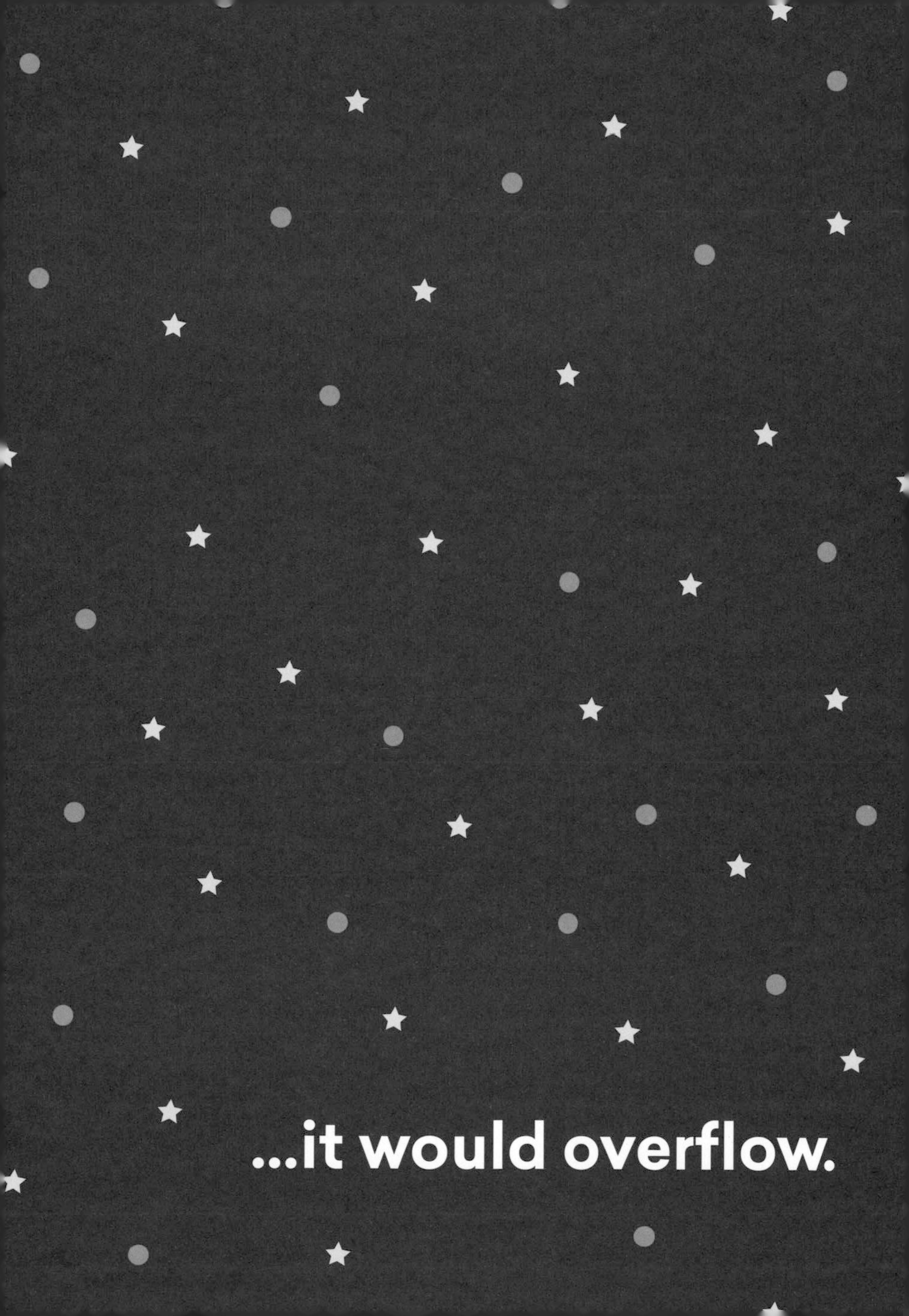
...it would overflow.

In a lifetime a person walks half way to the moon.

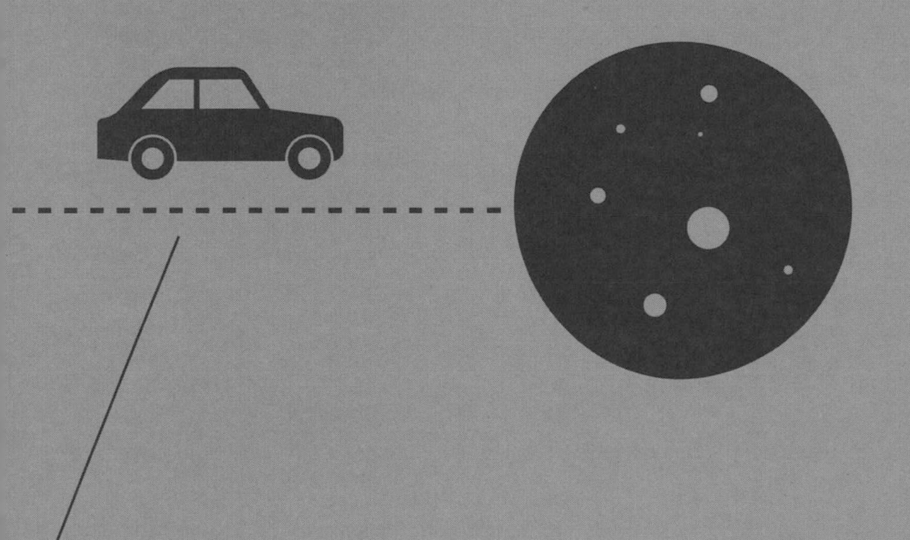

The average car also travels half way to the moon.

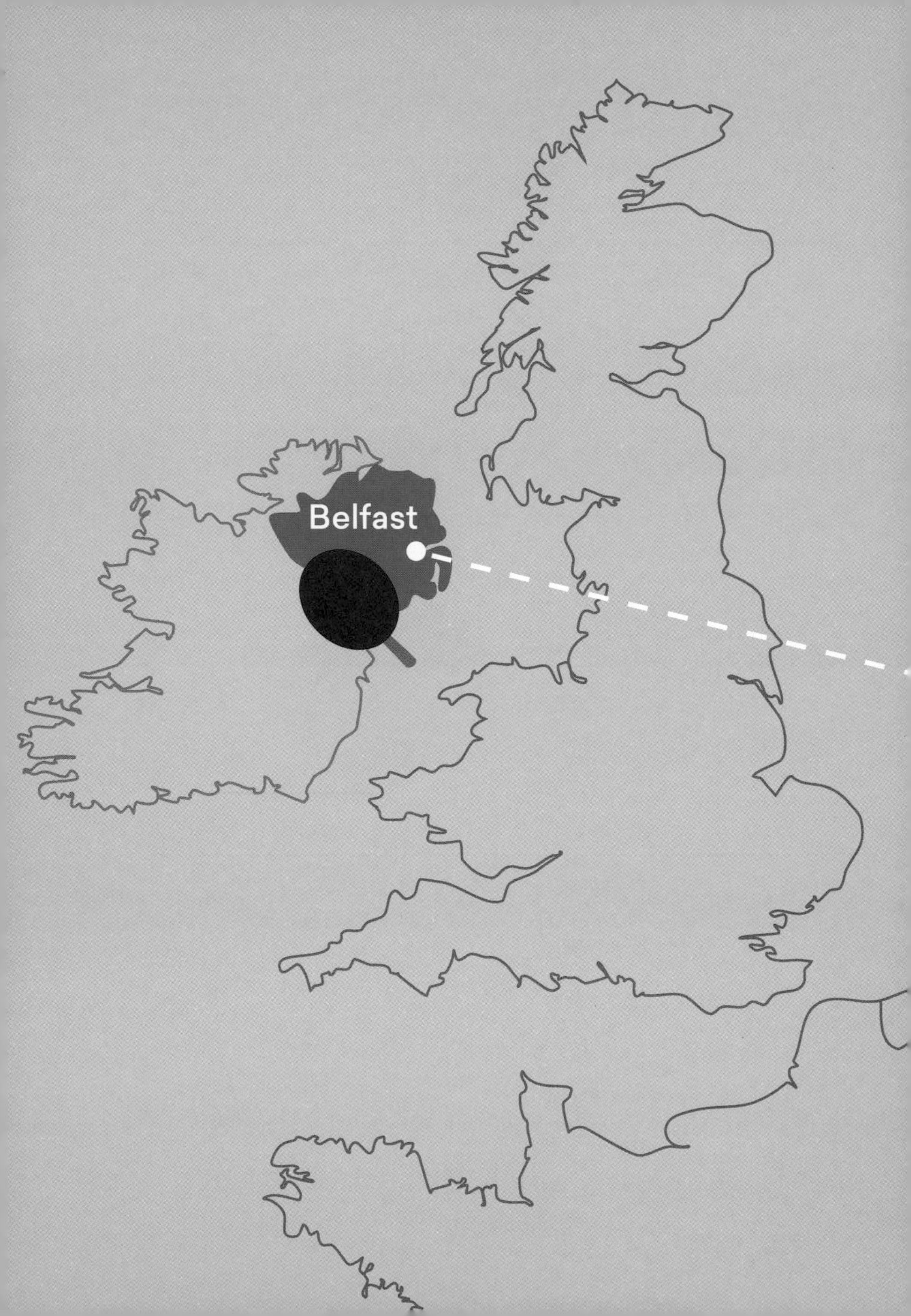
Belfast

If the Sun is a ping-pong ball in Belfast, Northern Ireland, its nearest neighbouring star, Alpha Centauri, is also ping-pong ball sized but in Berlin.

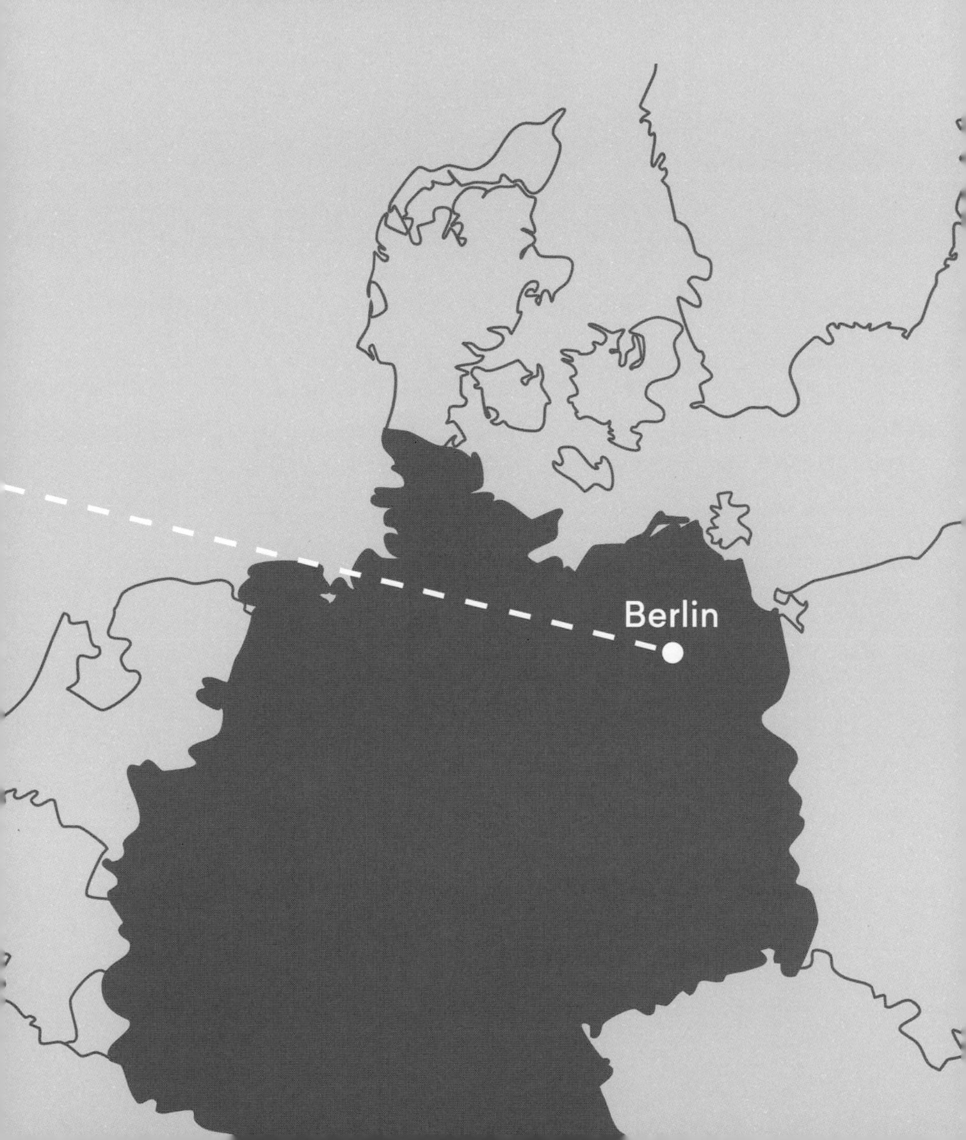

Rome

Attempting to pick up
an atom with your bare hands
would be like sitting
in a restaurant in Rome...

...trying to eat using
a fork gigantic enough
to reach a plate of spaghetti
on a table in Florence.

Florence

If the Earth became a black hole,

it would be smaller than a marble.

ECON

OMICS

Bill Gates earns so much money that if he drops a $100 bill...

...it does not make financial sense for him to spend time bending down to pick it up.

A **solid gold** Statue of Liberty and plinth would represent more gold than has ever been mined.

If American households allotted the same proportion of their weekly budget on food as Kenyans, they would spend $600 per person.

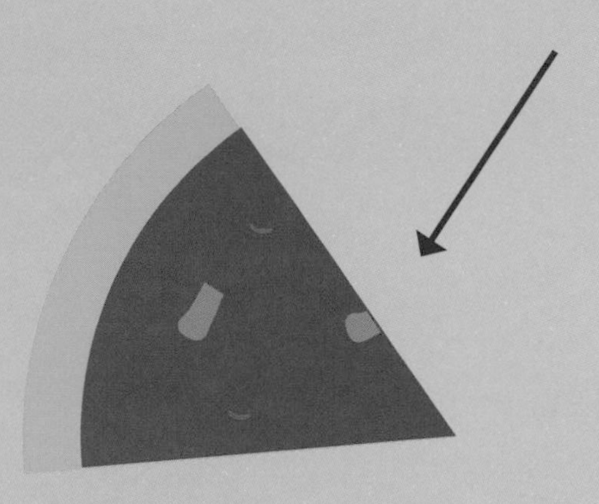

If Kenyan households allotted the same proportion of their weekly budget on food as Americans, they would spend $1 per person.

The annual return on illegal drugs would be roughly nineteen times greater than investing in legal pharmaceuticals.

If laptops had the same profit margin as perfume, they would cost the equivalent of:

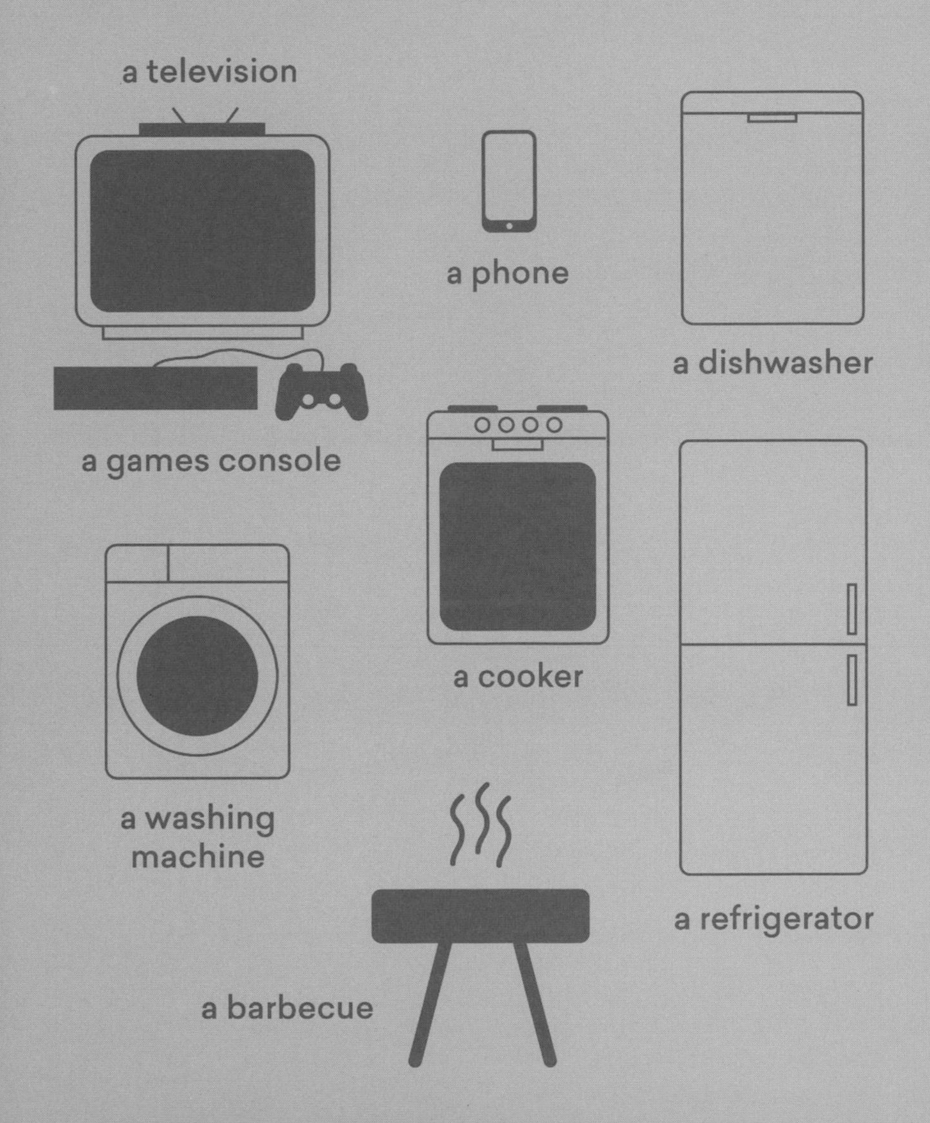

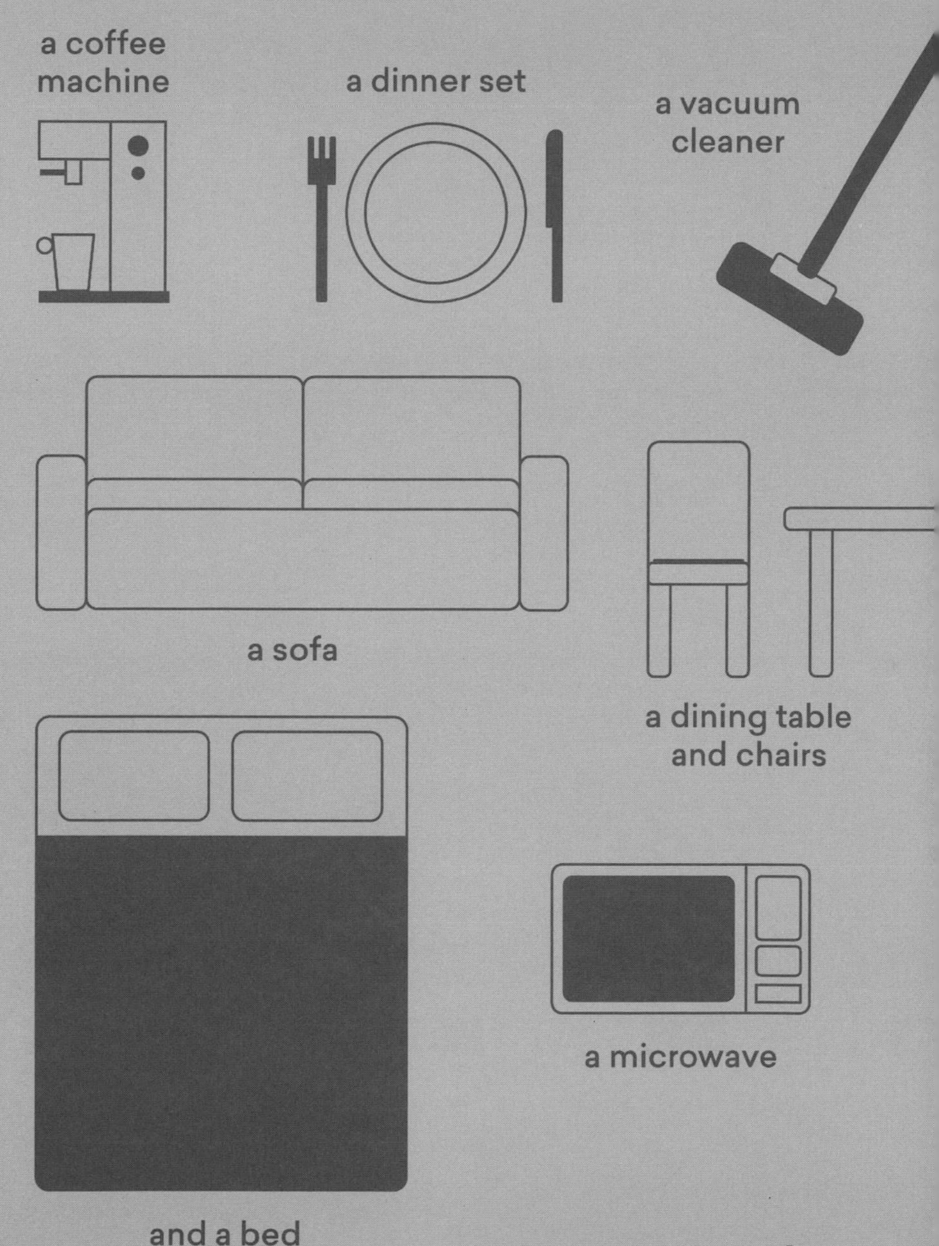

put together.

The value of Apollo mission moon rock is now so great, it would be cheaper to send an unmanned craft to bring more back.

Eight billionaire men own half the wealth of the world. They could ride around in a Limo on their half of the planet.

The rest of humanity would need a hundred million London buses in their hemisphere.

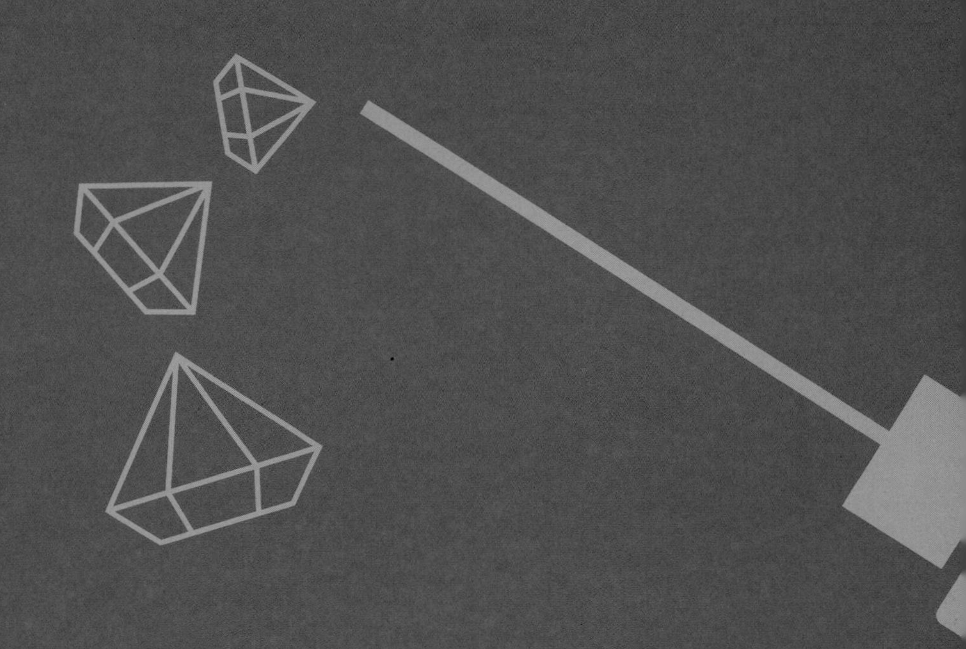

A kilogram of the most expensive substance ever made, Botulinum toxin (known in highly dilute form as Botox), would cost more than a diamond weighing a kilo.

It could also kill the entire human race.

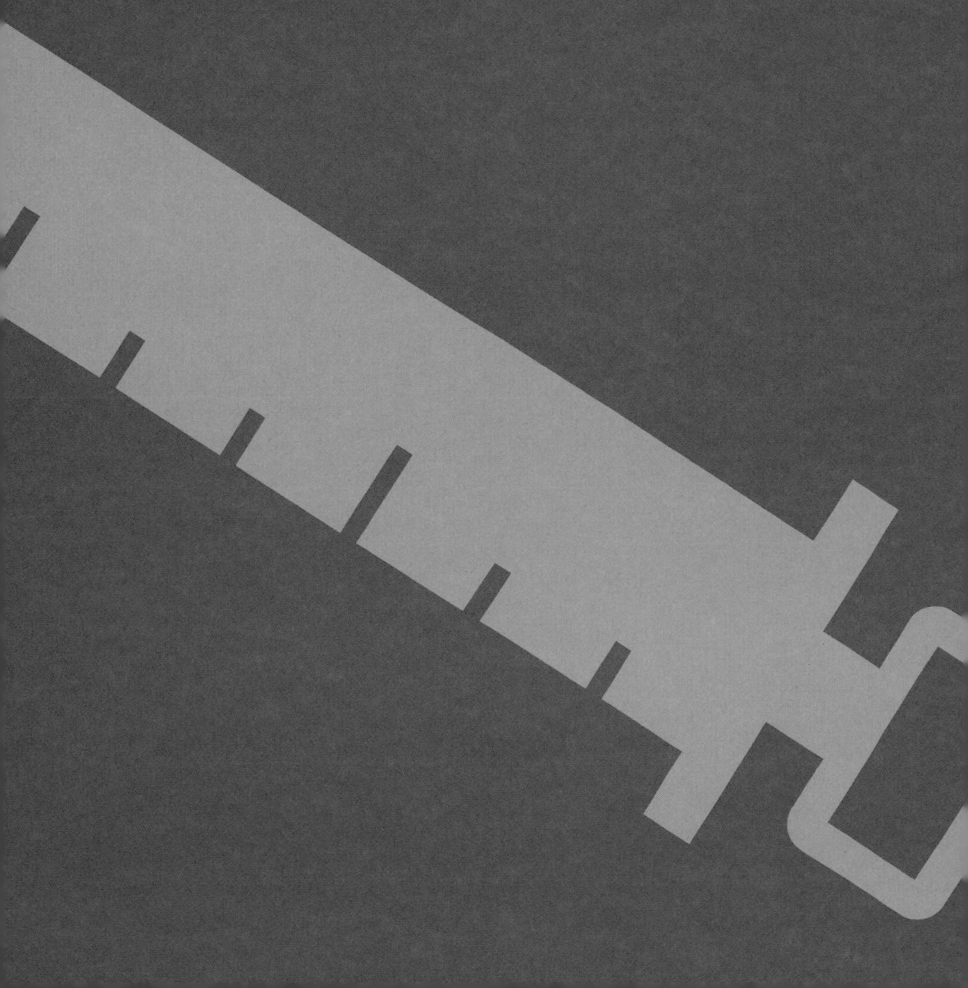

PO

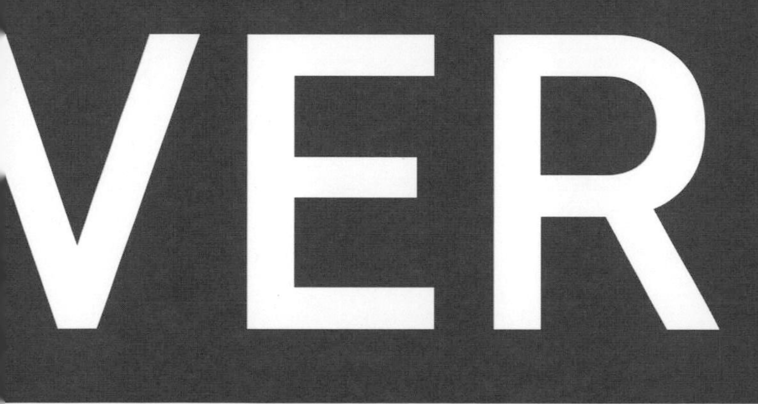
VER

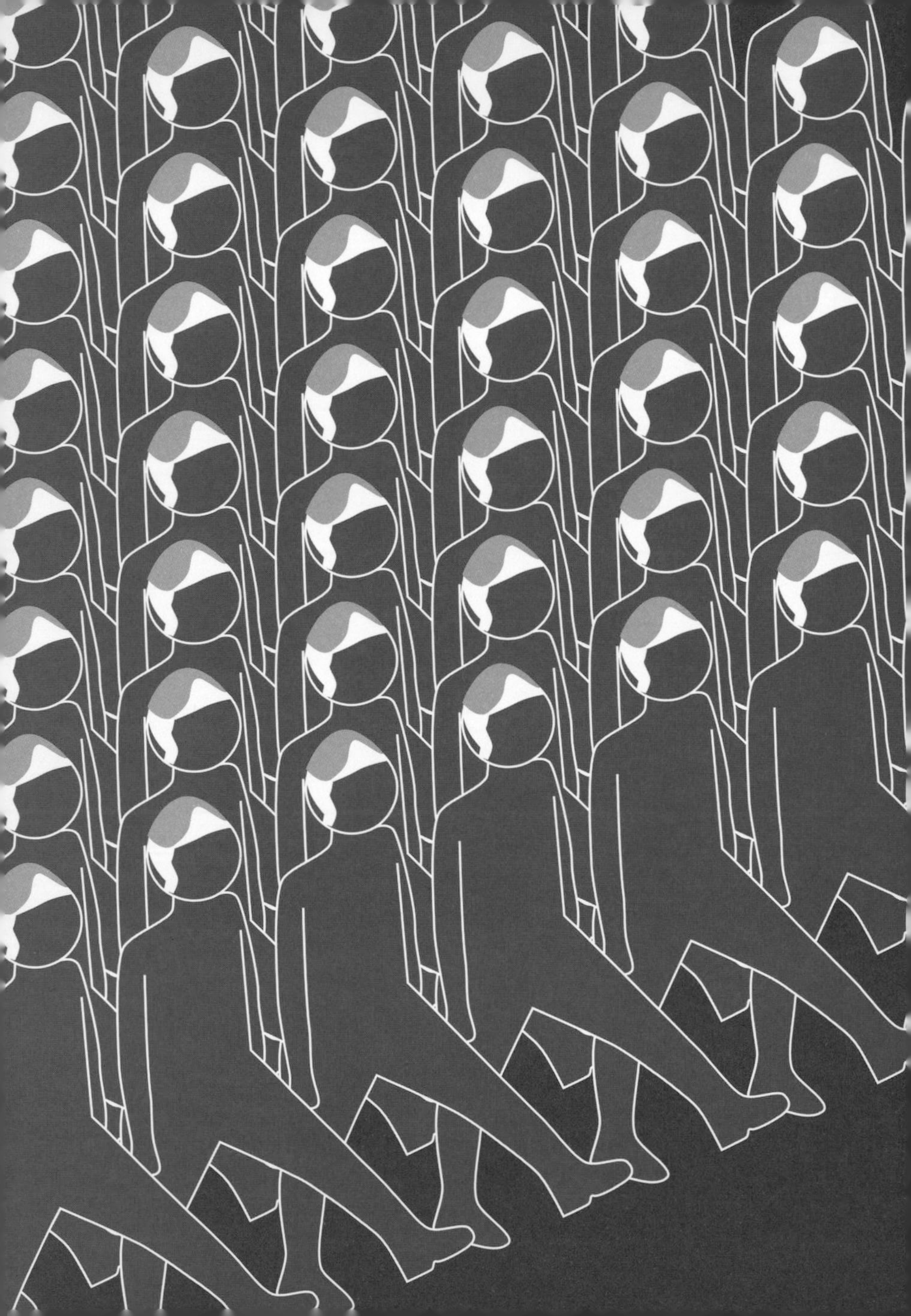

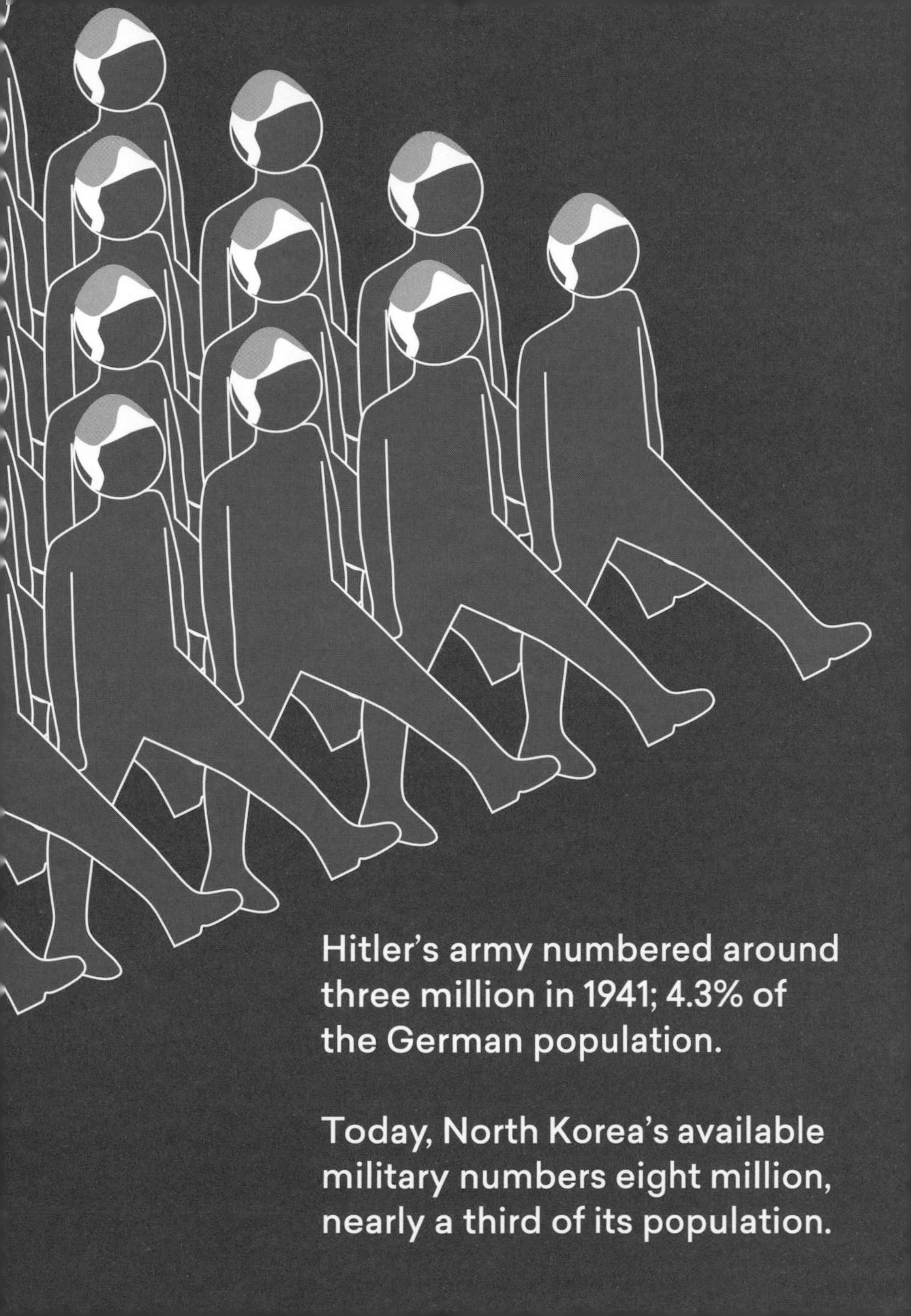

Hitler’s army numbered around three million in 1941; 4.3% of the German population.

Today, North Korea’s available military numbers eight million, nearly a third of its population.

A fifth of the world's total land mass is owned by seven kings, a queen, a sheik, a sultan, an emir and the pope.

In two hundred years, the Aztecs sacrificed about the same amount of people killed at Auschwitz over two years.

The Mongol leader **Genghis Khan** may have been responsible for the deaths of as

many as forty million people over his lifetime, reducing the

Years

Deaths

entire world population by as much as a tenth. Still five million less than **Mao Tse-tung** killed in the four years of his 'Great Leap Forward'.

The Uranium used in the 1945 atomic bomb 'Little Boy' which devastated Hiroshima, weighed the same as a Tic Tac.

The biggest nuclear weapon ever detonated, Russia's 'Tsar Bomba' in 1961, was 3,333.3 times more powerful than 'Little Boy'.

Human annual power consumption is equivalent to the energy Earth receives from the Sun in an hour and a half.

Scientology is worth over **$1.2 billion.** They claim we **each have two thousand** 'Body Thetan' alien spirits dwelling inside us.

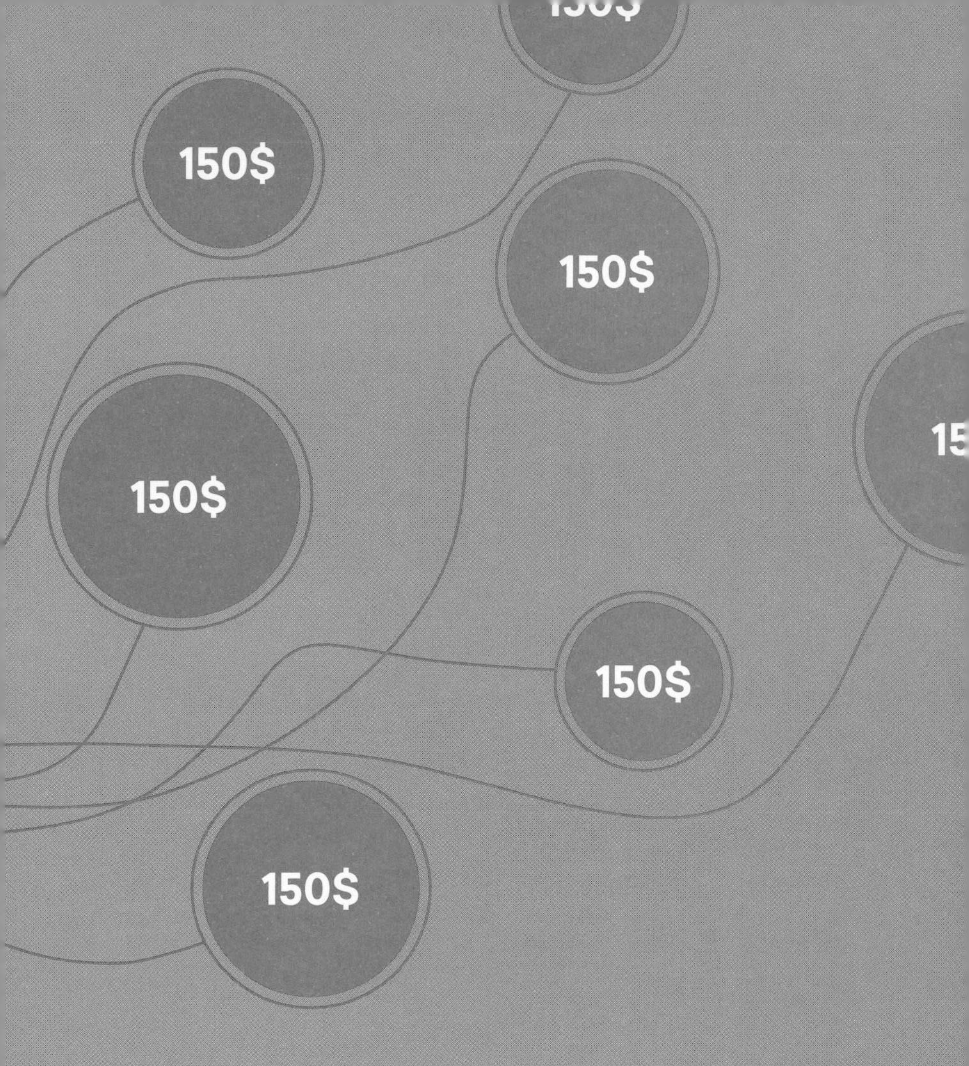

Each costs over **$150** to ‘clear’ generating potential global revenue of **$2.2 quadrillion**, thirty times all the money in the world.

ENVIRO

NMENT

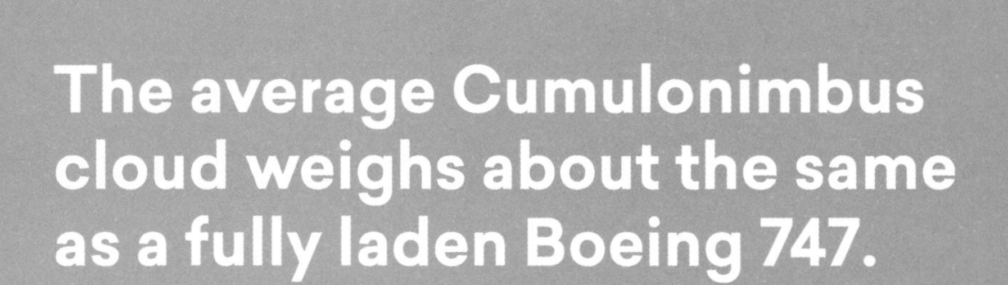

The average Cumulonimbus cloud weighs about the same as a fully laden Boeing 747.

At sixty miles per hour, driving eight hours a day every day,

it would take five lifetimes to drive all the roads in the world.

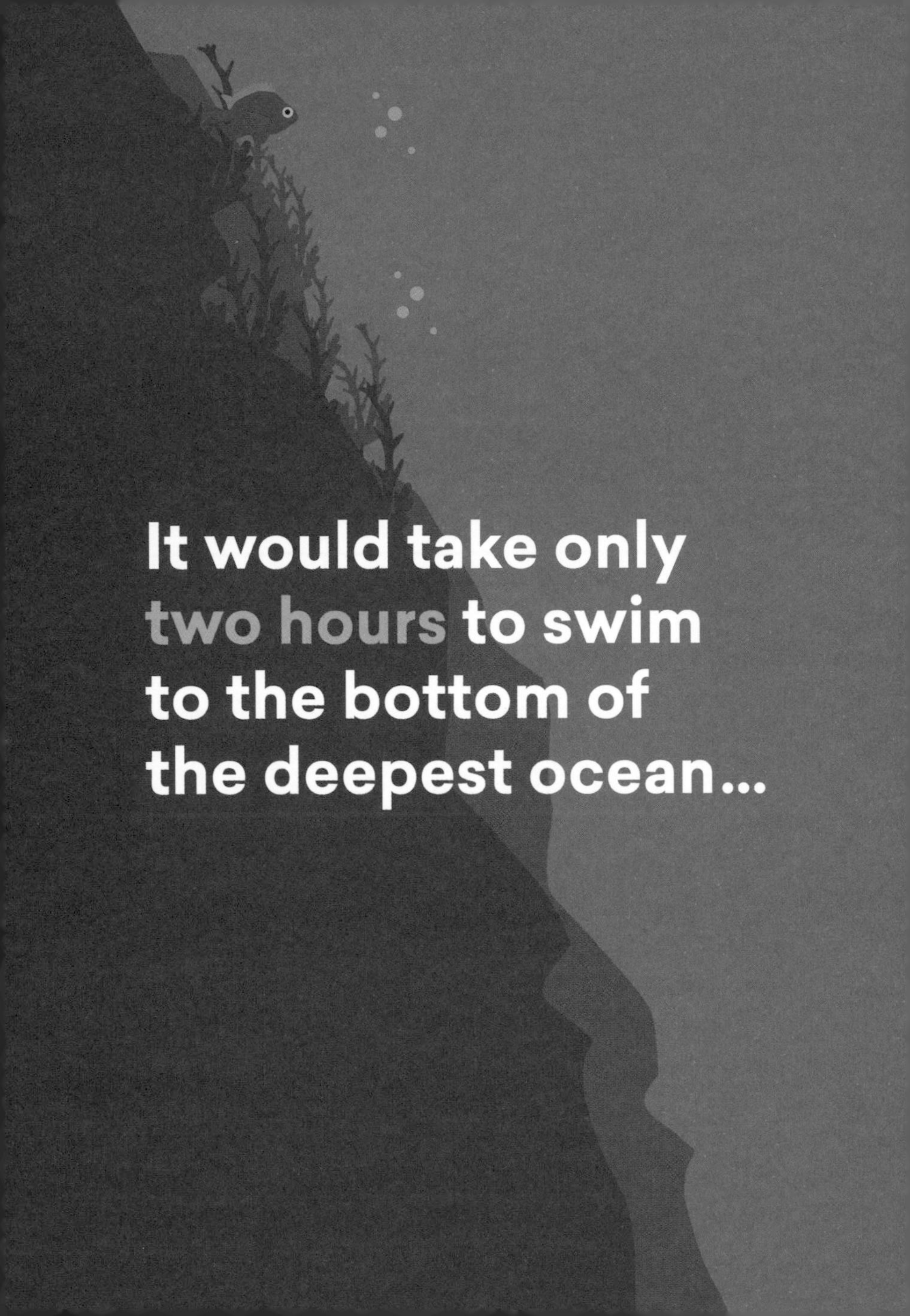

It would take only two hours to swim to the bottom of the deepest ocean...

...but the water pressure would compress you to the size of a tennis ball.

The great pyramid of Giza took **25,000** labourers and 20 years to construct.

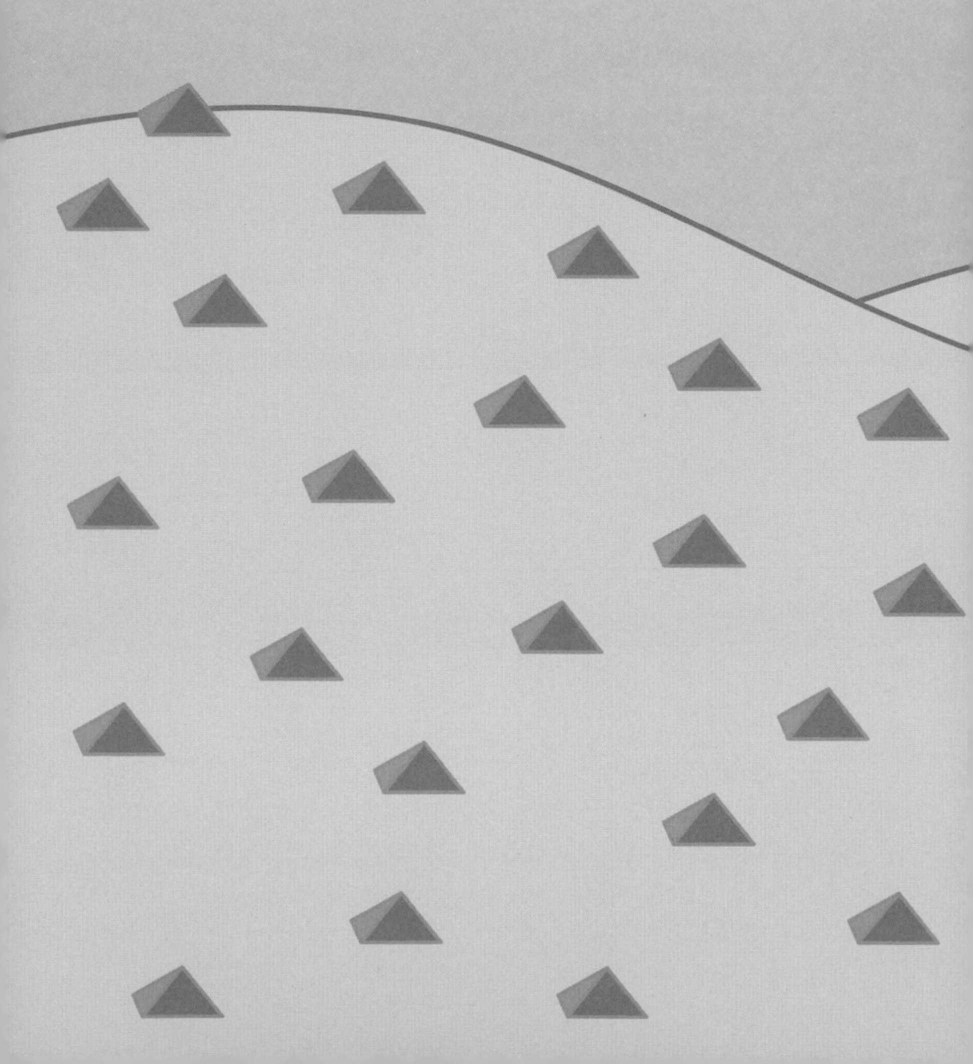

With modern techniques we could build 36 in the same time with half as many people.

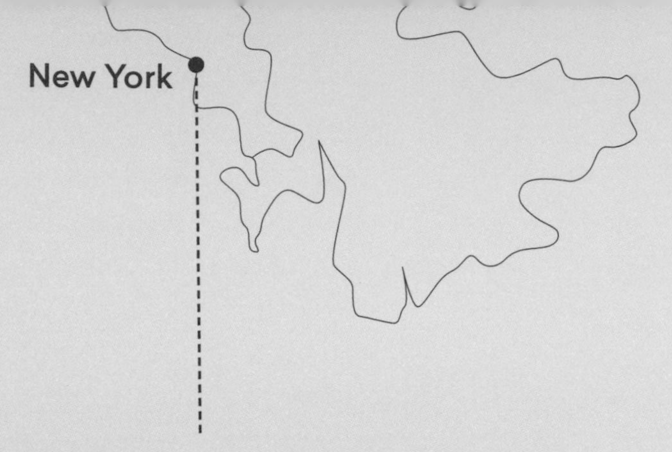

The distance to the centre of the Earth is 6,371km. Almost the same as between New York and Berlin. The deepest hole ever drilled is 12km deep; only half way from JFK airport to Manhattan.

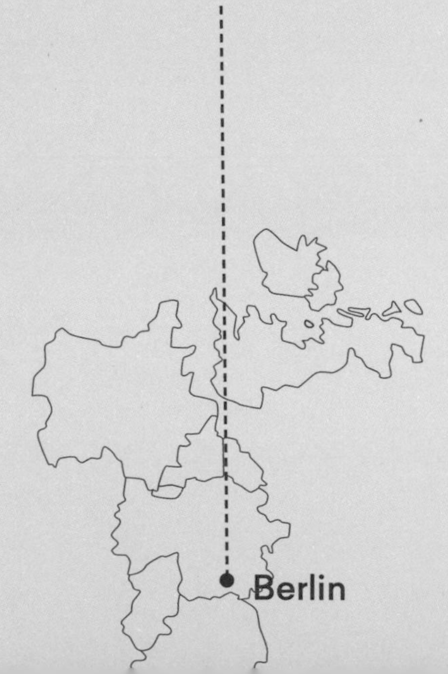

A ‘mega-colossal’ volcanic event, as is possible at Yellowstone Park in America, would erupt with more energy than five times all the nuclear warheads in the world detonating at the same time.

If Mount Everest were dropped into the Mariana Trench its peak would be underwater, leaving enough room to balance:

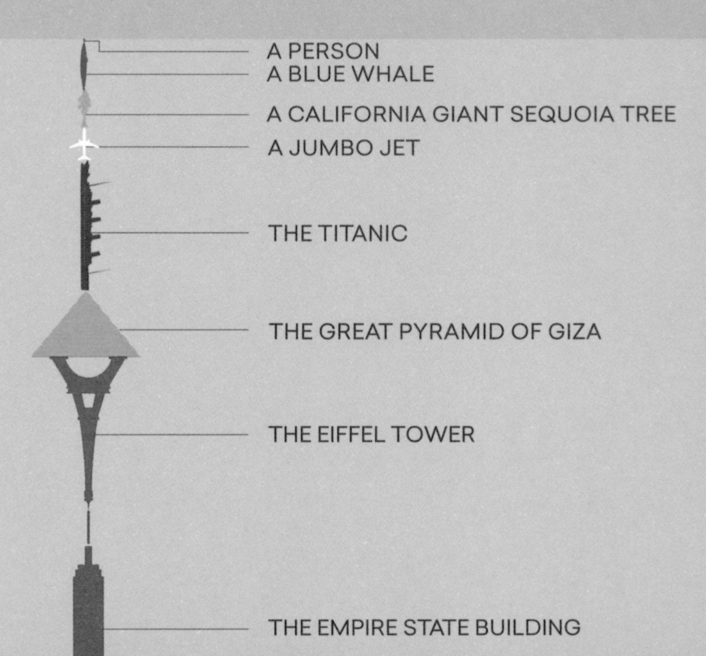

THE BURJ KHALIFA

The slowest tectonic plates of the earth's crust move at the same speed as your fingernails grow.

The fastest at the rate your hair grows.

TECHN

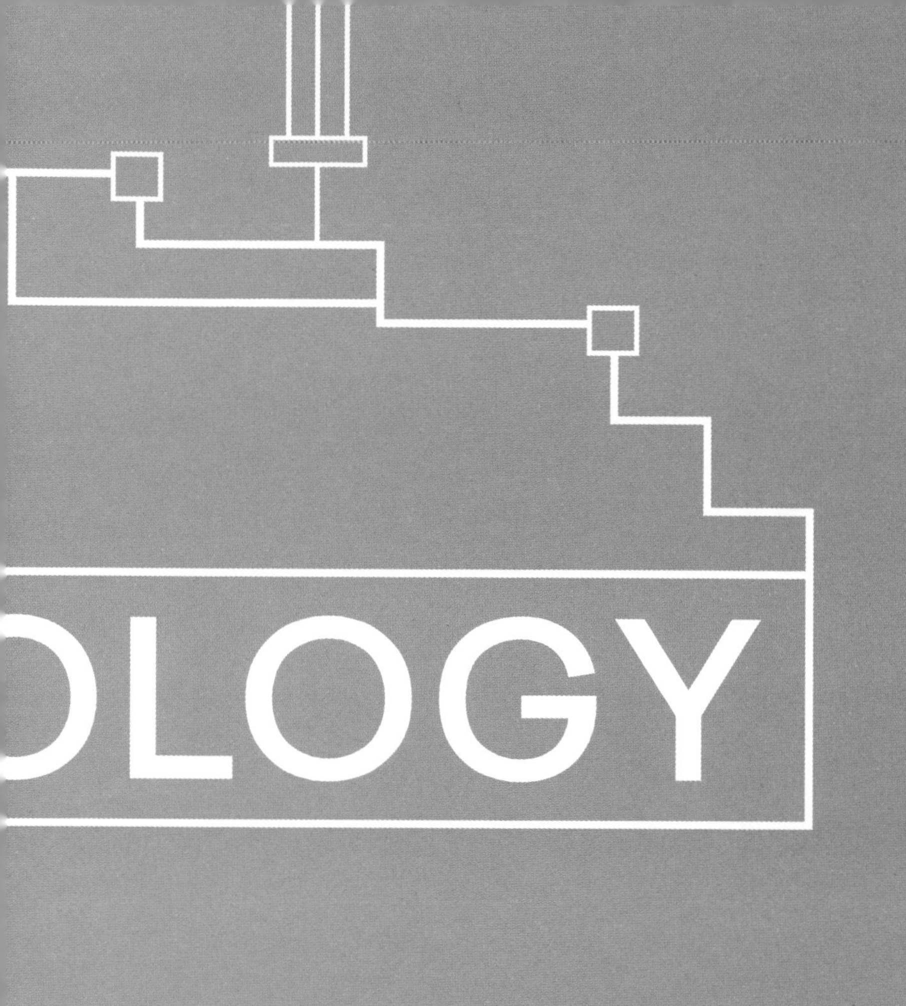
OLOGY

If you printed a days worth of Instagram photos and stacked them...

...it would be taller than Mount Everest.

As data itself does have slight mass, the entire internet weighs about the same as an egg.

Lego factories produce enough bricks to build a life-size human figure every minute.

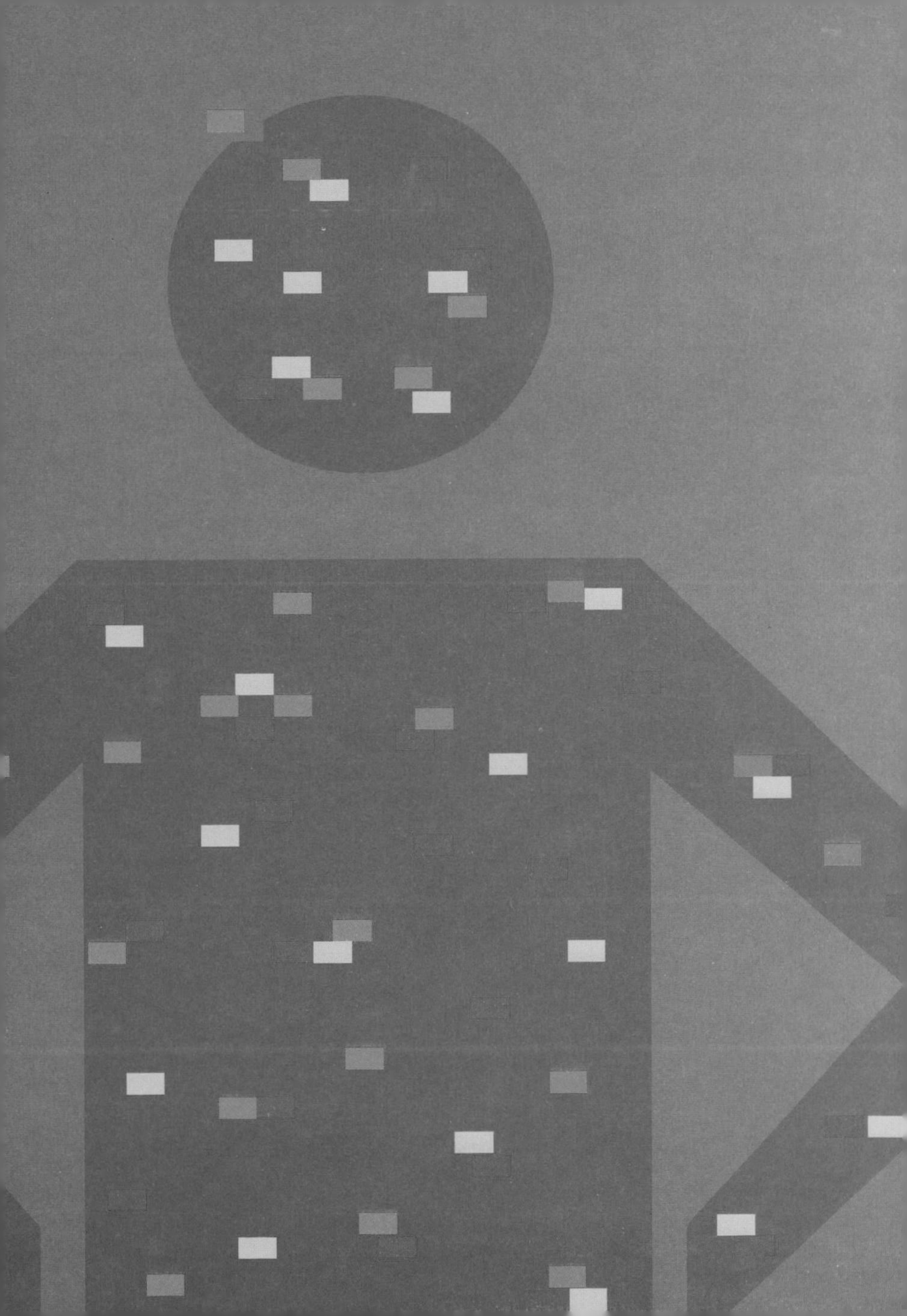

A copy of all the books ever published stacked on a single library shelf would stretch from Beijing to Bangkok.

Every page of the internet printed out would need a shelf nine times longer.

Every tenth of a second, we manufacture ten times more silicon chip transistors than there are stars in the Milky Way.

A Formula One car travels at the same speed as human thought.

If you google **'Google'**,
in about a second you will get more
results than there are human beings.

Google

7.5bn results in 1.0s

HUM

ANS

Over a lifetime, the average human sheds around two-thirds of their body weight in dead skin.

IN YOUR LIFETIME

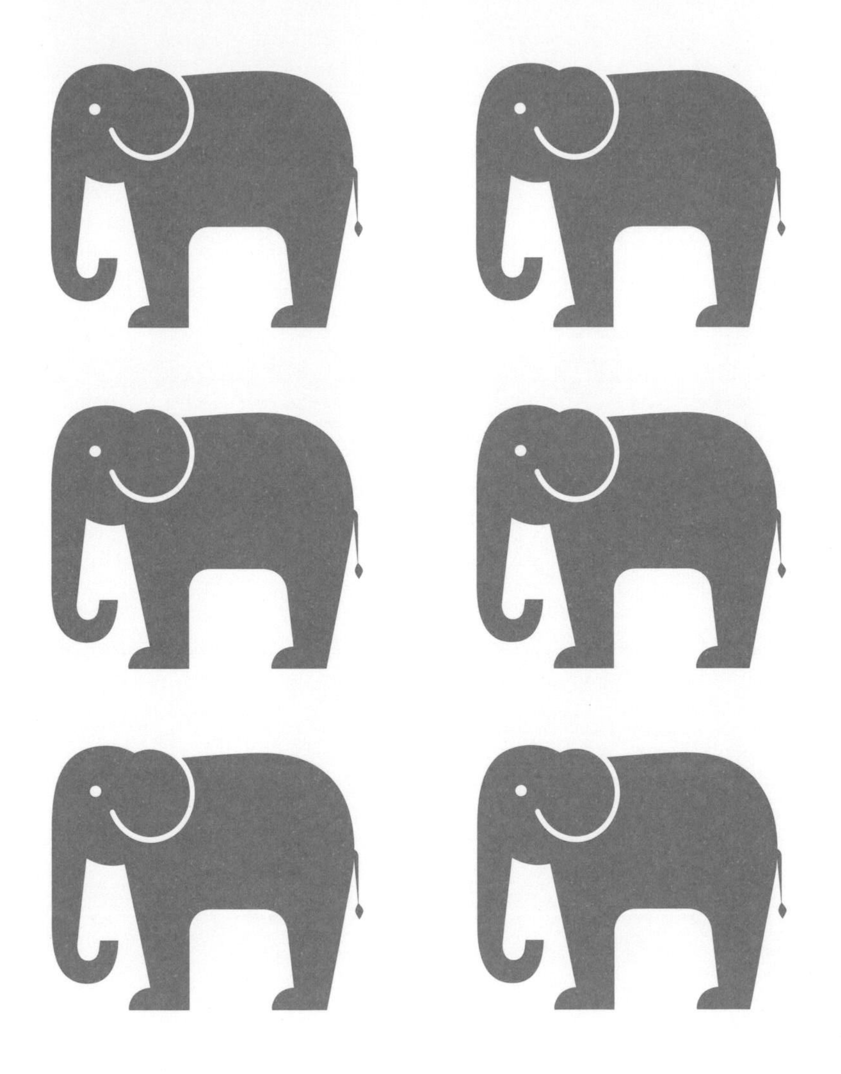

You will eat the equivalent weight of six elephants.

In your lifetime you will fill a domestic swimming pool full of urine...

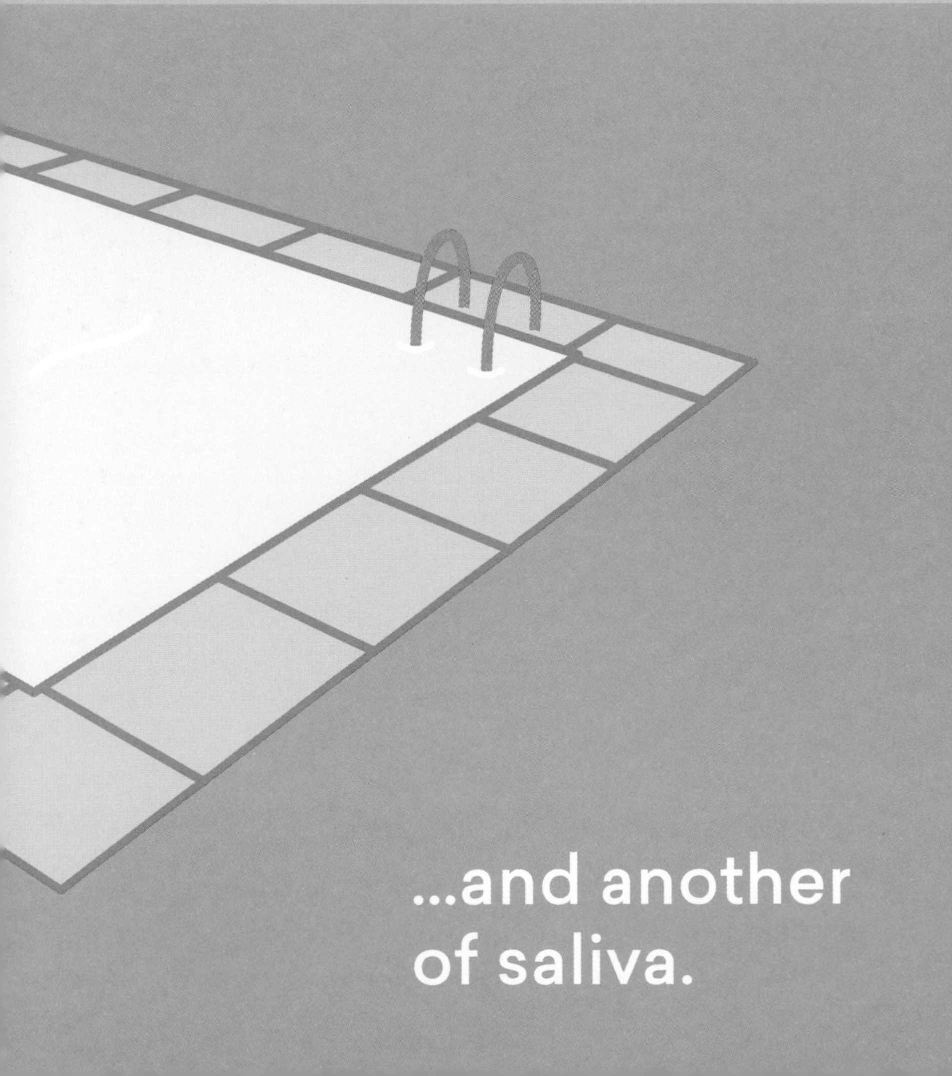
...and another
of saliva.

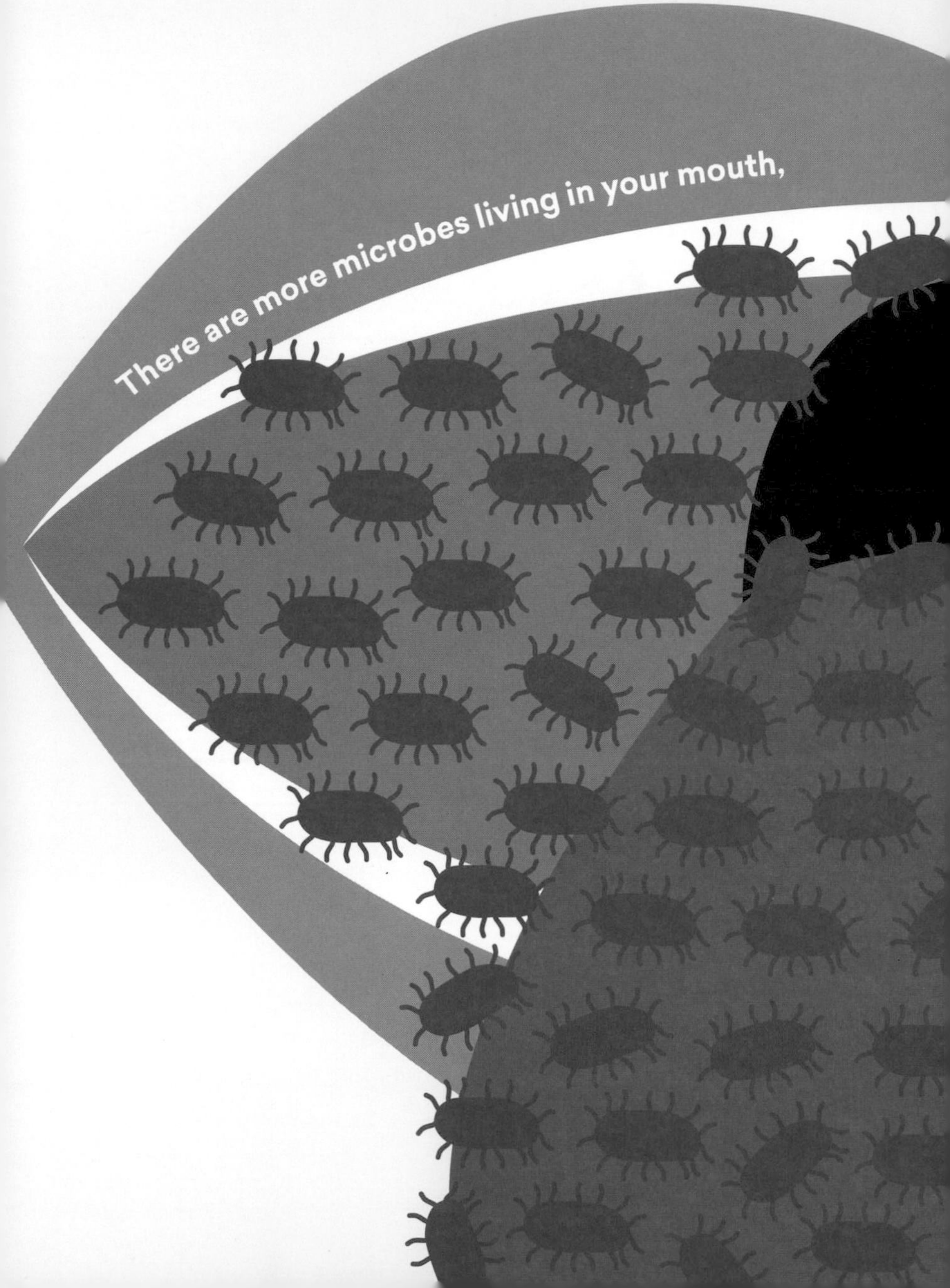

There are more microbes living in your mouth,

than there are people living on earth.

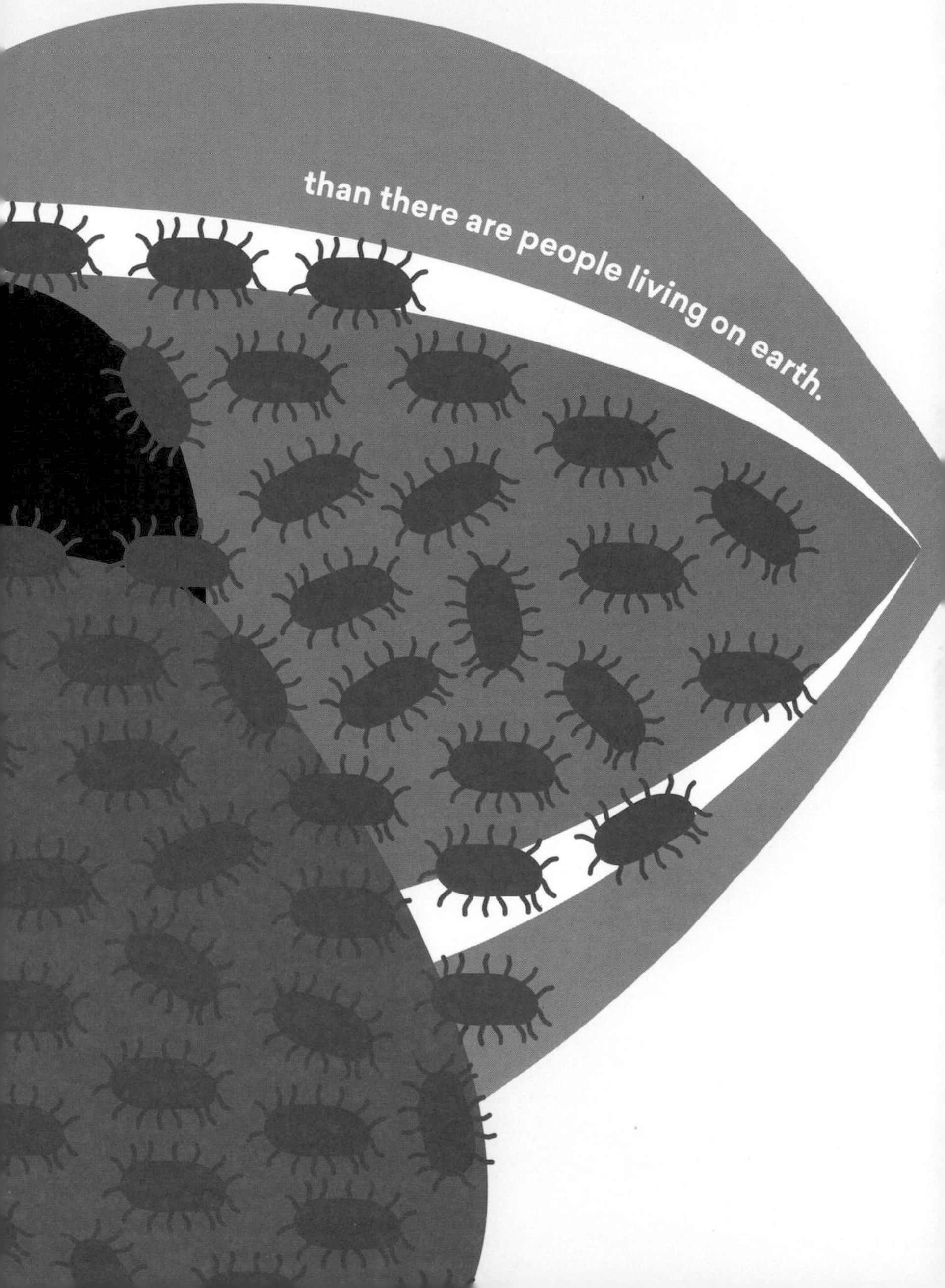

Humanity defecates at roughly the same volume per second as Niagara Falls.

If humans had babies at the rate rabbits do, our global population would double every three weeks.

The amount of crude oil at sea in all the world's supertankers equals the amount humanity is overweight.

ANIM

MALS

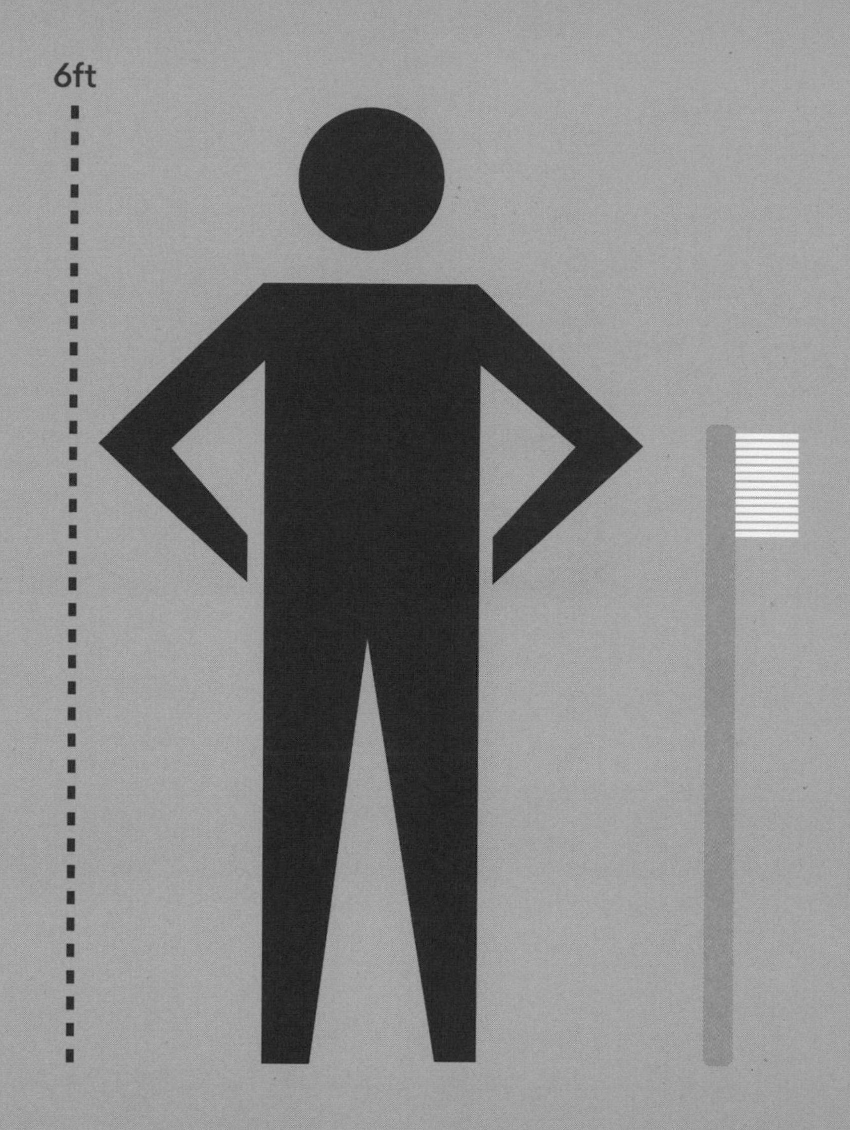
6ft

Comparing their dental size to ours, an elephant's toothbrush would be four feet long.

T. Rex's arms were only slightly longer than an average humans but his shoe size was around

100

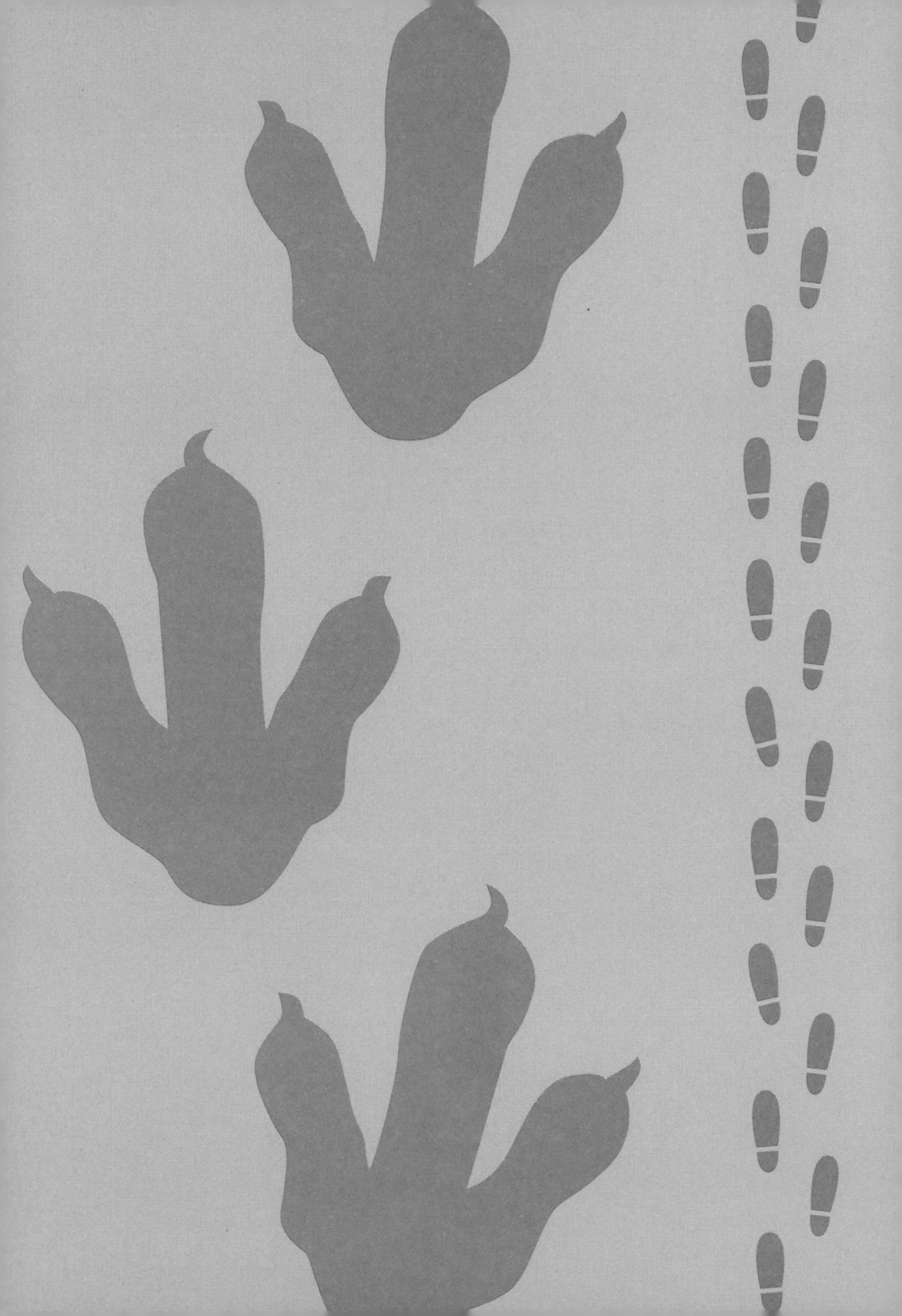

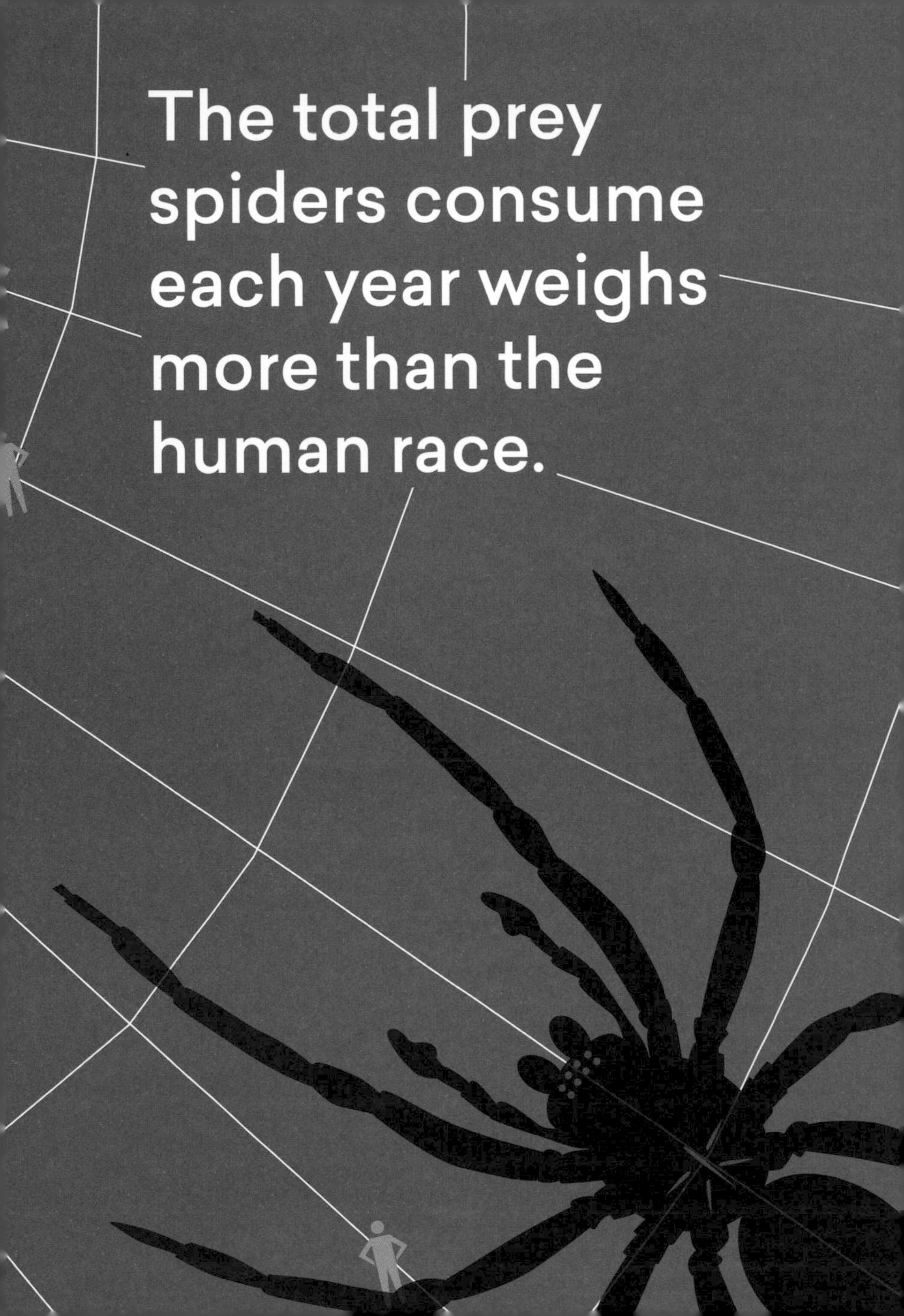
The total prey spiders consume each year weighs more than the human race.

The most famous internet cat and dog have more Facebook page 'likes' than there are pet cats and dogs in the world.

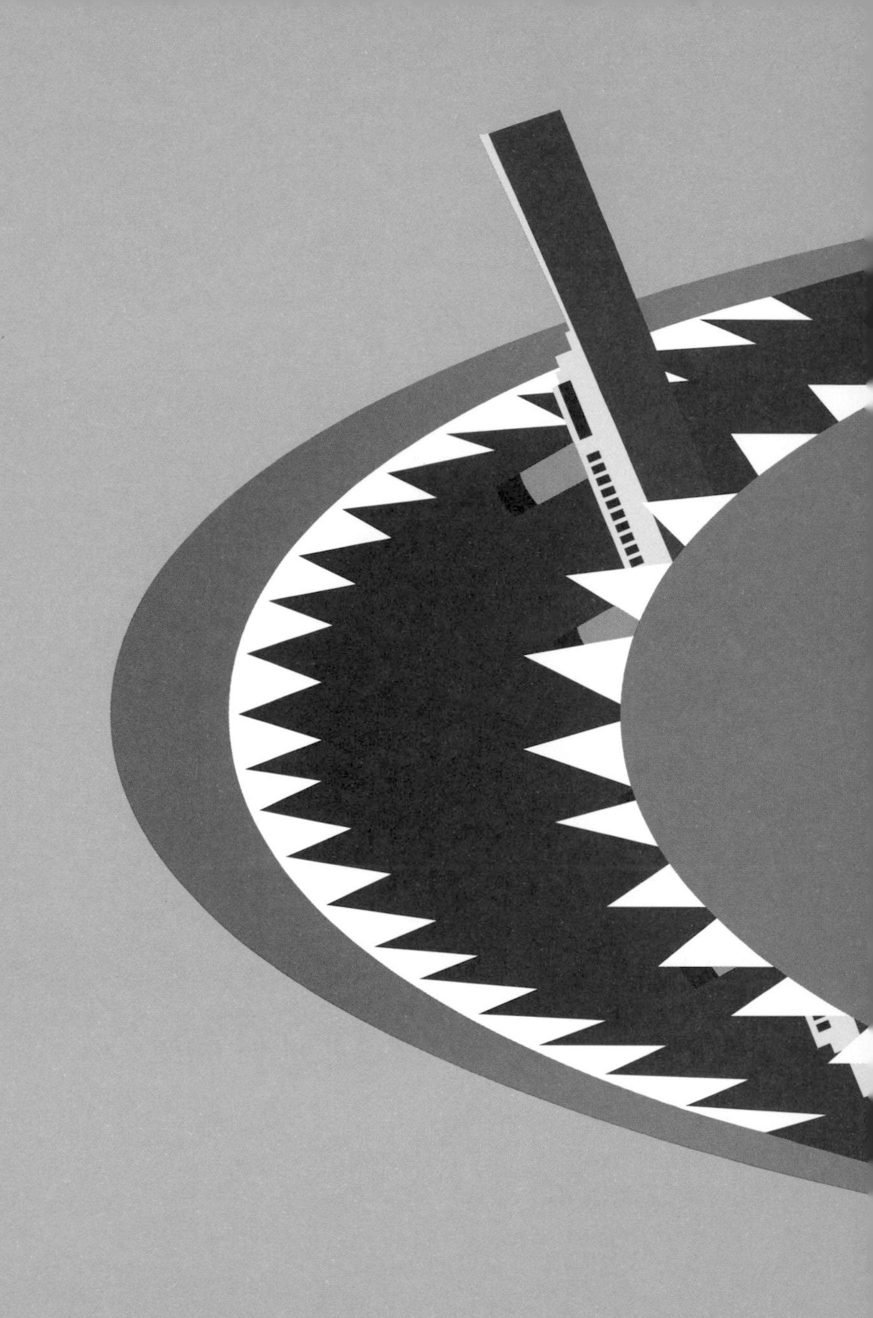

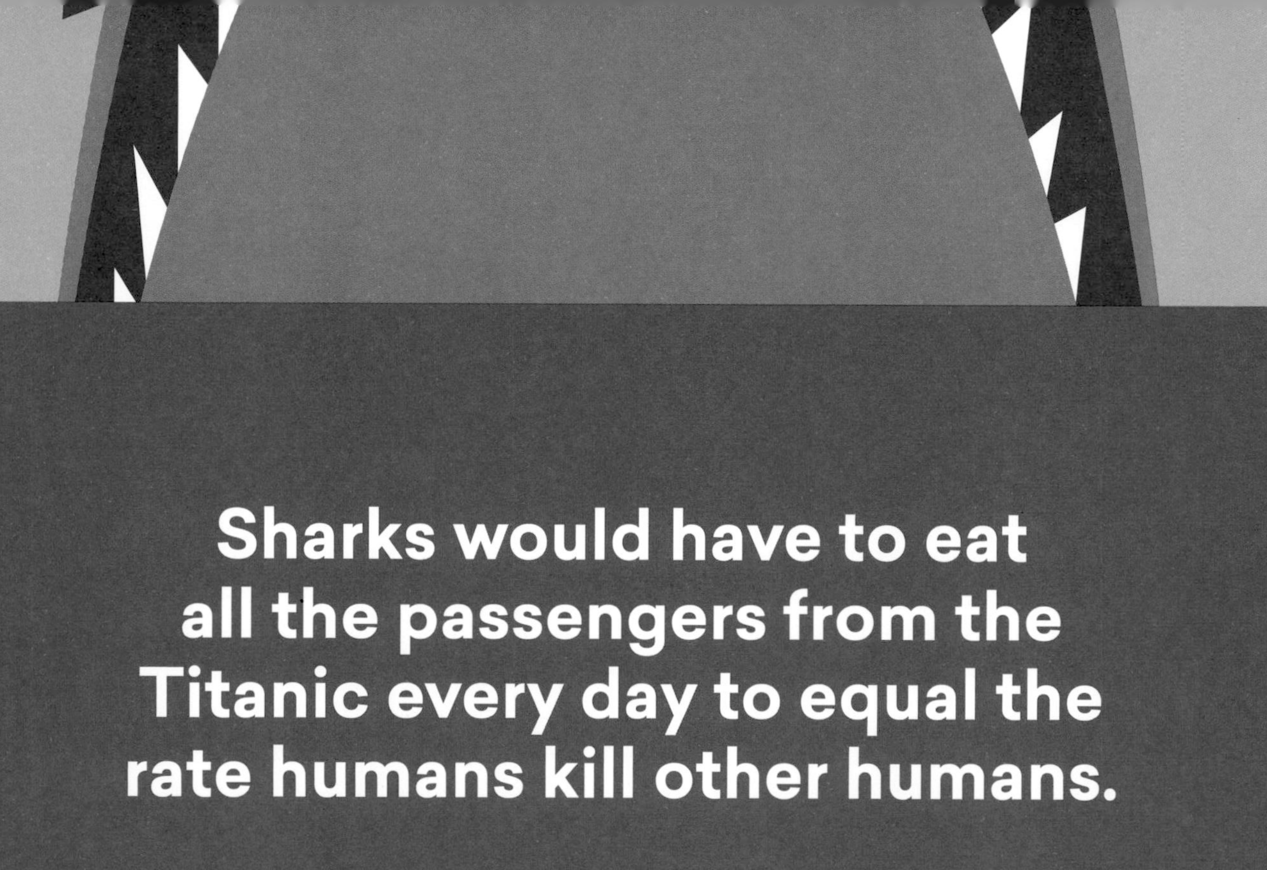

Sharks would have to eat all the passengers from the Titanic every day to equal the rate humans kill other humans.

If humans have ruled
the earth for 5 minutes...

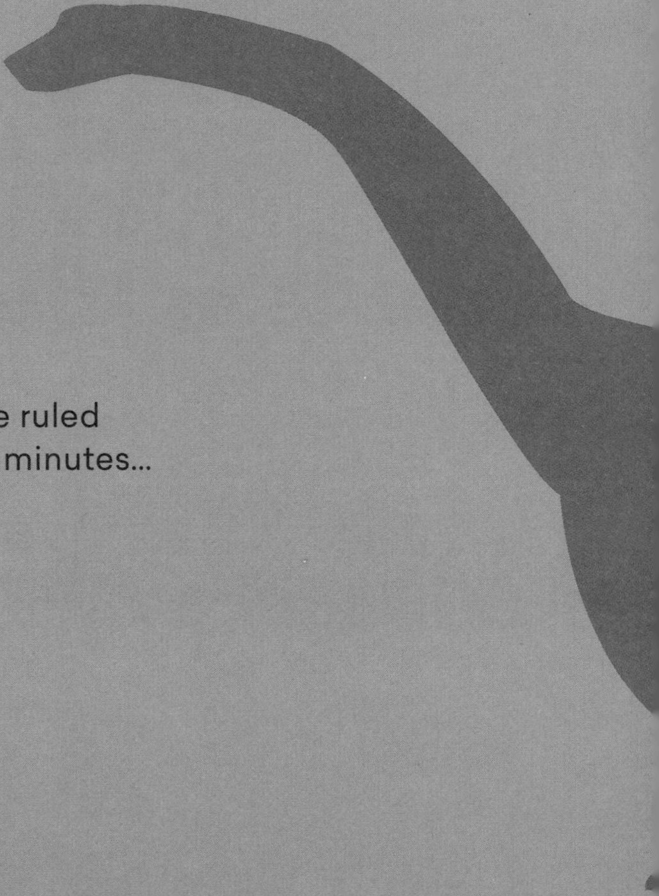

...the dinosaurs were in charge for around 3 months.

Human-sized ants would hardly be able to move, but humans shrunk to ant size would be twice as strong as them.

If your sense of smell were scaled to that of a dog's...

...upon meeting a friend you would know everything they had touched that day, who they had seen, where they had eaten, and what mood they were in.

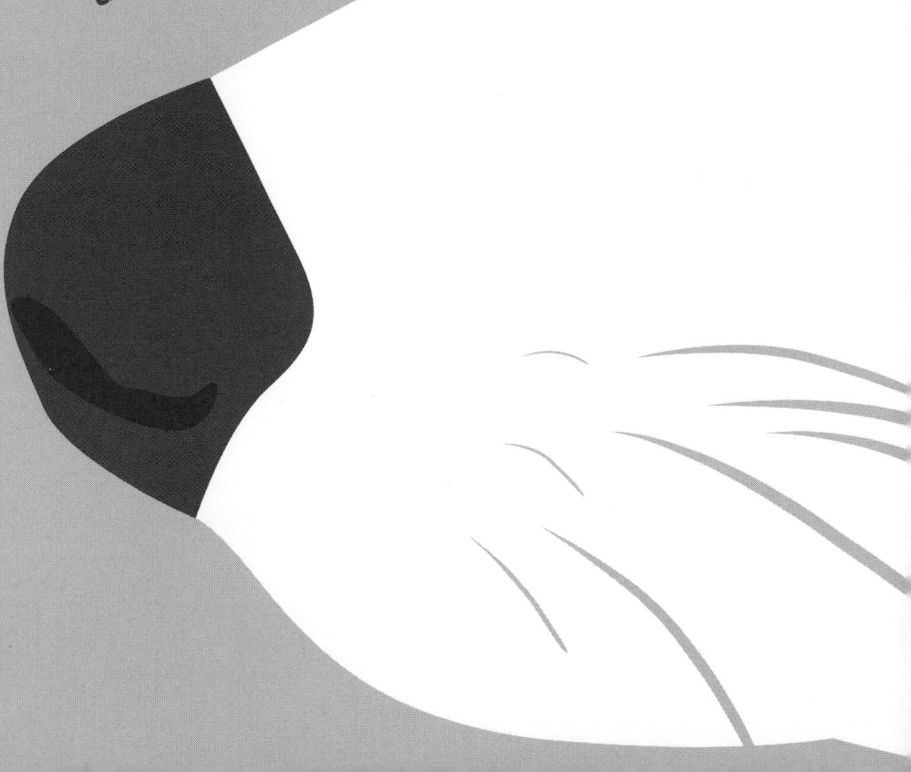

MIKE FAIRBRASS

Mike Fairbrass is a writer and creative consultant. The former Head of Modelmaking and Photography at Richard Rogers' practice, he has exhibited at The Royal Academy, The Pompidou Centre, Paris, MOMA, New York, The Beijing Capital Museum and in The Architectural Biennale, Venice. He has written for *The Architect's Journal*, taught at The Royal College of Art and run educational workshops at The Design Museum, The V&A and The Royal Academy of Arts in London. "Mystic" Mike is also the Virgin Radio UK official (satirical) astrologer.

DAVID TANGUY

David Tanguy is Creative Director of the London design studio Praline, which he founded in 2000. He has created numerous successful projects worldwide, encompassing graphic design to branding, exhibitions and art direction. David has worked with clients including Tate, the Royal Academy of Arts, Barbican, Harvard GSD, V&A and many others. He has designed numerous books for international publishers and art galleries, and his work has been recognised with several international awards. He is the co-author of the books *Unforgotten New York* (Prestel, 2015) and *Pop City – New York* (Editions du Chêne, 2016).